Luath Press

The Scot and his (

As a boy, Wallace Lockhart, experienced the bothy life on Angus farms. As a young man, he entered the grain trade. He has grown oats and harvested them, stooked them and stacked them, milled them and baked them, eaten them and drunk them and researched their history.

This little book tells much about the oat crop. It also tells much about the Scots and how the oat has shaped their lives.

G.W. Lockhart

THE SCOT AND HIS OATS

Luath Press

To Dod Beattie,
Smiddyhill, Stracathro,
Angus,
whose knowledge of crops and animals
makes me realise how little I know.

Single Furrow Plough

PREFACE

One of the great pleasures I have derived from the writing of this little book has been the opportunity it has given me to meet and discuss the oat crop and its products with so many people who share my interest and enthusiasm for the traditional Scottish scene. For the knowledge they have imparted to me, I am extremely grateful.

In particular, I must express my thanks to Mr. T. Paterson of Rutherglen, Mr. J. T. Lambie of Paterson's Oatcakes, Livingston, and Mr. James Walker of 'Walkers of Aberlour' for their very considerable help. To Mr. A.J. Munro of Atholl Estates I am indebted for keeping me right on Athole Brose. As ever, the staff of Angus Libraries and Museums spared no pains to be helpful, and I acknowledge the permission given by their Director, Mr. G.N. Drummond, to use photographic material. Likewise, for encouragement and permission to reproduce photographs, I am in the debt of Dr. Alexander Fenton and staff of the National Museum of Antiquities of Scotland, who do so much to keep our heritage alive. For allowing me to draw on material, I am obliged to Mr. Wilfred Taylor, and for permision to include individual photographs, appreciation is expressed to Mrs. M. L. Mercer, Honorary Curator of Fife Folk Museum, Ceres, Mr. Bruce Walker A.R.I.B.A. of Dundee and Lerwick Public Library, Shetland.

Finally, for capturing the spirit of what I have tried to convey in this book, I am immeasurably in the debt of the well-known Border artist, Jim Coltman of Hawick, for his splendid sketches.

G.W. Lockhart
Linlithgow

ACKNOWLEDGEMENTS

For permission to use photographs, thanks are gratefully extended to the following:-

Lerwick Public Library, Lerwick, Shetland.
Mrs. Mercer, Ladybank, Fife.
Bruce Walker A.R.I.B.A., West Ferry, Dundee.
James N. Walker,Strathspey Bakery, Aberlour.
T.W. Paterson, Burnside, Rutherglen.
National Museum of Antiquities of Scotland.
Angus Museums and Libraries, Forfar.

THE SCOT & HIS OATS

CONTENTS

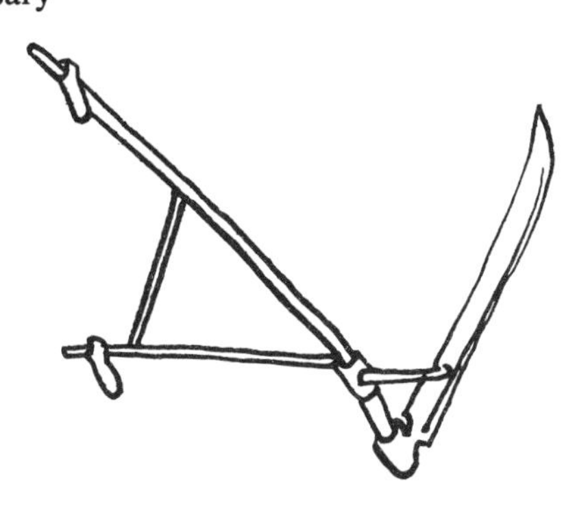

Stooking Oats

INTRODUCTION

As a boy, I hated porridge. The grey, bubbling mass in the black pot did not hold for me the relish it held for others. Worse than that, it seemed to me, in my little world, that everything centred around the making and eating of it. The Howe of Strathmore was filled with field upon field of oats, all apparently destined to finish up in the breakfast bowls of small boys. How that gently swaying, delicate, golden crop which so tingled the senses could bring such misery in its train was something my mind could not absorb. But I was the odd man out. My friends loved their porridge and oatmeal in all its forms. As we cooked our trout, taken from the burns in the Angus Glens, the fish were invisible under their smothering of oatmeal. My pride would not let me complain, and I joined in the acclaim that we were feeding like kings.

The corn fields at harvest time were sheer bliss. The stubble crunched under our feet as we ran round the stooks and if a mass of bites was the price to pay for being able to hide under the sheaves while stalking a courting couple, it was of little importance.

The passing years brought greater knowledge of the oat crop. One learned to identify the different varieties by rubbing out a few grains of the crop in the hand: '*Marvellous*' with its green tip, tubby '*Onward*' and '*Golden Rain*' with its distinctive colour. Wild oats were seldom seen, but when they were noticed standing proud of the crop they were ruthlessly rogued out. Scots Seed Oats were famous and no contamination was permissible. Occasionally the Potato Oat was encountered. Legend has it that this oat was discovered growing in a field of potatoes around the year 1800 and, because of its plumpness, it was kept for seed and multiplied. Its original habitat was the subject of much correspondence to farming papers in the early nineteenth century, claims being made for Turkey and South America as well as Essex and Cumberland. What is certain is that it was superior to any other oat at the time of its introduction, and that in the Fiars' Courts (which are referred to later in this book), it was for many years separately priced.

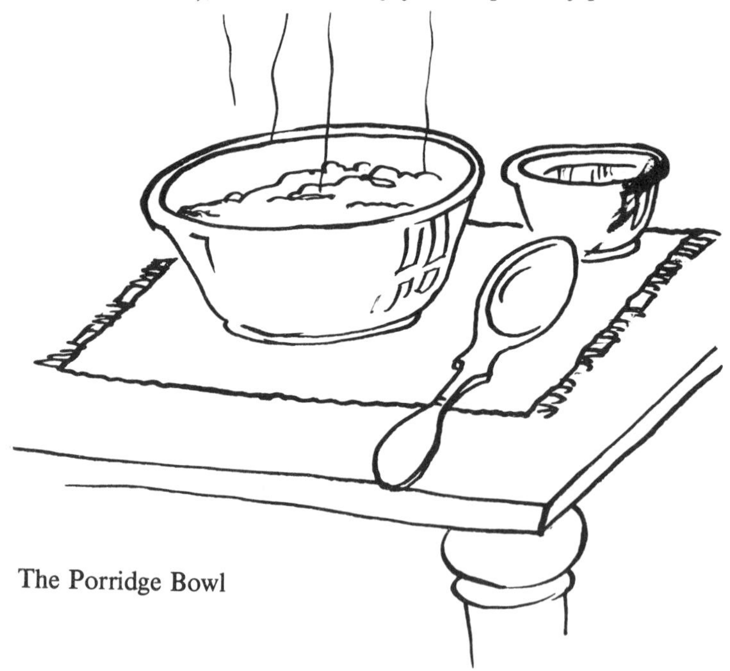

The Porridge Bowl

Army service followed school, and this intervention provided an escape from porridge. As everyone knows, the Scot makes porridge with salt; the cooks in my regiment, being English, made it with sugar. Thus, proclaiming that I would not offend my nationality, I was able to avoid the fearsome stuff with dignity, even, I fear, with a touch of rudeness. *'Haggis'* might have been my soldiering nickname, but eat sweetened porridge I would not.

The passing years (or was it army sustenance?) civilised my palate, and when I entered the grain trade I found I could nibble away at the freshly ground oatmeal issuing from the mill with some enjoyment. A later sojourn in Canada introduced me to the coating of porridge with maple syrup and the much more acceptable honey. But it was the waving crops of oats in British Columbia and the Prairie Provinces which confirmed my affection for the crop, although their yields and plumpness of grain were not on a par with Scottish crops. Every year, at the Winter Agricultural Fair held in Toronto, a competition was held to find the world's best sample of oats, and every year, it seemed, the challenge was won by oats submitted from a Scottish farm. I returned to this country a dedicated oatmeal addict.

The Cottar's Kitchen

The nationalism of the Scot expands when he leaves his native land. He needs no lessons in propaganda. Hogmanay, St. Andrew's night, Burns Suppers, ensure the world knows when the Scots are around. And the menus, with their proliferation of haggis, oatcakes, Athole Brose and crannachan remind three thousand million people of the staple foods of a nation. The wonder of it is that the Scot is not only tolerated, but encouraged and emulated in his vauntings. No other race would get away with it. Wilfred Taylor, that doyen of Scottish journalists, in his book '*Scot Free*', recounts the story of the Scot who arrived in some far off jungle land to discover, to his horror, that St. Andrew's day had never been celebrated in the locality. He immediately started issuing invitations. Out of a total neighbouring population of 360 Europeans, 250 attended. An Englishman who had accepted his invitation was so stung that he set about arranging a St. George's Day the following year. It was attended by seven people, all of them Englishmen.

Hugh MacDiarmid, of course, took a more trenchant view of such Scots activities, and in '*A Drunk Man Looks At The Thistle*', rasps out:

'You canna gang to a Burns supper even
Wi'oot some wizened scrunt o' a knock-knee
Chinee turns roon to say, "Him Haggis — velly goot!"
and ten to wan the piper is a Cockney.'

Sadly, it must be admitted that some exiles' celebrations are more noted for their enthusiasm than their authenticity. Arriving to speak at a St. Andrew's Night dinner in Yorkshire some years ago, I was handed a large glass of Athole Brose by my host, who expressed the hope that it would get me into the right frame of mind. It was an ambiguous invitation. The Brose had the consistency of putty.

In this little book I have tried to trace the journey of the oat crop; not only from field to factory but over the centuries, as an integral part of Scottish life. Such a historical and social study, superficial as it may be, inevitably shows just how much the Scot owes to his oats. I hope you will enjoy the journey with me.

THE OAT CROP

'Hear, Land o' Cakes, and brither Scots,
Frae Maidenkirk to Johnny Groat's.'

Burns.

'Oh, soldiers! for your ain dear sakes,
For Scotland's, *alias* Land o' Cakes.'

Fergusson.

'Till eke their cheer ane subcharge furth sho brocht,
Ane plate of groatis, and ane dish full of meal.'

Henryson.

From at least the fifteenth century poets and chroniclers have regularly told of the importance of the oat crop in Scotland. It is perfectly understandable; seldom has a crop been so associated with a people and their way of life. For hundreds of years, oats, in one form or another, served as a staple diet of the population. Together with its straw, the oat helped to fatten the famous black cattle, while the straw on its own was plaited into curtains for doors and used for paper making. Oatmeal had value as a currency in the payment for rent or wages or to form part of a dowry. Oat poultices were used to draw poisons from the body, and the harvesting of the crop provided themes for many songs, especially in later years for the great bothy ballads of the North-East.

Soldiering has always been a popular Scottish vocation, and in the Middle Ages Scottish soldiers were renowned for their ability to travel quickly and fight lustily on an apparently meagre diet. Froissart, the great European traveller and historian of the fourteenth century, comments on this fact in his writings:

> 'Under the flap of his saddle, each man carries a broad plate of metal; behind the saddle a little bag of oatmeal. They place this plate over the fire, mix with water their oatmeal, and when the plate is heated, they put a little of the paste upon it, and make a thin cake, like a cracknel or biscuit, which they eat to warm their stomachs.'

The Porridge Pot

Thus is set out the first recipe for oatcakes. The relationship between the Scots fighting man and his oatcakes is obviously one that continued over the centuries, for we read that a patriotic Highland lady, upon learning of the defeat of Bonnie Prince Charlie's army by the Duke of Cumberland at Culloden, set up a roadside stall providing fresh oatmeal bannocks for men making their escape from the battlefield. It seems the Scot may be caricatured for his meanness or love of whisky or haggis, but when it comes to his meal more than a note of seriousness is evident.

Oats of course, with their high energy value, are the great food for horses. As every Scottish schoolboy used to know, Dr. Johnson commented that the Scots lived on the food which in England was given to horses, and indeed went so far as to give such a definition in his first dictionary. Perhaps it was the riposte:— *'And where will you find such men and such horses?'* from Lord Elibank that led to its later suppression. Fergusson, that great vernacular Scottish poet and fore-runner of Robert Burns, was obviously stung by Johnson's remark, and in his poem describing the treat given to the eminent literary figure by the professors of St. Andrews University, was determined to set the record straight:

'Mind ye what Sam, the lying loun,
Has in his Dictionar laid down?
That aits, in England, are a feast
To cow and horse, and sicken beast;
While in Scots ground this growth was common
To gust the gab o' man and woman.
Tak tent, ye regents! then, and hear
My list o' gudely hameil gear,
Sic as hae aften rax'd the wame
O' blyther fallows mony time:
Mair hardy, souple, steeve and swank,
Than ever stood on Samy's shank.'

There is, however on record an event which acknowledges the importance of the oat being an acceptable food for both man and horse. During the Boer War, the beleaguered garrison of Mafeking was able to keep itself alive by making a gruel from the contents of the horse feed boxes.

The oat is a very old cereal crop. Believed to have originated in Asia, it spread westwards and was known on the European continent more than two thousand years ago. Hippocrates (5th century B.C.) refers to the crop, and Virgil (1st century B.C.) was apparently enough of a farmer to complain that oats had the ability to choke out barley; and to the Greeks goes the honour of first recognising the value of a porridge made from oats.

Although the oat can be grown under widely varying conditions of soil and climate, it is at its best where the weather is cool and early summer rainfall provides the conditions for the kernel to fill out with slow maturity. Scotland provides such conditions and the crop's journey north was therefore assured. With natural selection being aided by the husbandman's selection of the best plants for seed, yields of grain increased and the crop began to take an increasingly important place in the economy. By the thirteenth century it was established as a popular food crop sharing a place with barley. Improvements to the oat crop continued to be made, and by the early 1600's oats were classified in the Fiars' Courts, where prices were agreed for oats and oatmeal. Not a great deal of information is available about the origins of Fiars' Courts, but they appear to have had the function of relating the rent, and in some cases the feuduty, to be paid by the tenant to the value of the crops produced on the farm. Until the Church of Scotland rationalised the stipends paid to ministers, it was customary to base the stipend on the value of the crops grown in the parish, using the Fiars' Courts prices.

By the eighteenth century, white, black and grey oats were recognised as distinctive and the naming of strains within these categories was soon to follow. As the grain trade developed in the nineteenth century, seed from various continental varieties was imported for crossing with indigenous strains, leading in turn to the breeding and growing of higher yielding strains. The selection of seed became more scientific, and leading farmers, seed merchants and the Colleges of Agriculture began to establish field trials to ascertain the best and most appropriate varieties of oats to grow in different parts of the country. All this led to the magnificent crops of oats seen this century, although at the present time, due to market forces, barley is the predominant cereal crop grown in Scotland.

It had long been considered that seed produced in the harsher climate of the northern parts of the United Kingdom was healthier, plumper and higher yielding than that produced in more southern climes. This was the basis on which was established the seed export trade to England, which, even to the middle of this century, saw thousands of tons of seed oats every year make the journey south by road or rail. For a while, too, a modest export trade was carried on with some Southern African countries. At one time some race-horse owners contended that black oats gave their steeds extra pace, but that specialist trade, unfortunately, has decreased, if not disappeared altogether.

As the crop grew in popularity in Scotland, so too did the sophistication of the tools and equipment asociated with it. The jump from the picking of the ears by hand to the introduction of the sickle was indeed a gigantic technological advance. The sickle, which frequently had a toothed cutting edge, allowed a man, or just as often a woman, to cut a good quarter of an acre of oats in a day. It is strange that the scythe, which had been known in Roman times, and was presumably used for mowing grass, took such a long time to be employed in the harvesting of grain crops. A factor might be that, because the oat harvest was later in the Middle Ages than it is nowadays with earlier ripening varieties of oats, the crop would have been laid by the weather and thus difficult to scythe. Certainly the first scythes to be used on a large scale had a cradle arrangement fitted to them, which allowed the crop to be set out in more regular swathes, and the rate of cutting per man rose to over an acre a day.

Up to the beginning of the nineteenth century, gangs of blue-bonneted Highlandmen and Highland lasses moved to the Lowlands every year to take part in the harvest, gradually working their

Harvesting Team

way north again as the harvest progressed. By the mid-nineteenth century, this work had largely been taken over by the Irish, who would arrive by boat on the West coast, each man carrying his own scythe and the leather apron which would be worn when working.

The tying of the cut stalks was done by women and followers. In those parts of the country where communal farming was practised, sheaves were tied in individual ways as an aid to identification. Stooking was the next step in the harvesting process and this entailed standing a number of sheaves in two lines against each other. The oats would stand for two weeks or so in the stook to allow ripening to complete and the grain to dry out. Local custom decided the number of sheaves in a stook and this could range from six to fourteen, smaller stooks being more common in the wetter parts of the country. Finally, the sheaves would be taken to the stackyard which was normally situated near the barn, although a windy spot had advantages as it allowed the drying out of the grain to continue. Where horses or ponies were used to do this leading, as the movement to the stackyard was termed, a '*kepper*' would be fitted into the animal's mouth to prevent it from eating the sheaves.

Stacks were built round a tripod and on a raised wooden or stone floor to allow some circulation of air and make it more difficult for rats and mice to take up residence. The sheaves were laid circular fashion with the heads of grain innermost. The diameter of the stack was anything up to fifteen feet and its height might exceed twenty feet. The stack builder worked round and round on his knees as the sheaves were fed up to him. The last few feet of the stack were tapered, and the final operation was to thatch the stack to make it waterproof and rope it down to withstand heavy winds. Frequently a little dolly or peerie decoration made from straw was placed on the crown of the stack.

It must be remembered that in the days before the advent of the combine harvester, the harvesting of the crop involved long hours of toil spread over many weeks. The end of the harvest was an occasion for great satisfaction and rejoicing and was marked by a festivity known as a *Kirn,* or a *Meal and Ale,* or a *Muckle Supper.* In the North-East, the term 'Meal and Ale' was used to describe the drink provided as well as the occasion itself; and a potent mixture it was too, consisting of whisky, ale, sugar and oatmeal. Stories abound about the happenings at Meal and Ales, and have been passed down the generations, particularly in bothy ballads. Fiddles and melodeons would play; there would be drinking, dancing and singing, leading to wilder activities such as riding pigs bare-back round the steading. Rituals also centered round the last sheaf to be cut. It might be given to the best milk cow in the byre, or shared out among the beasts at Christmas, or given to the first horse to start the next year's cultivations.

The first methods of threshing the sheaves to separate the grain from the straw were extremely primitive and involved hand rubbing the heads or treading the sheaves by oxen — and that method at least goes back to Biblical times. One of the first pieces of equipment to be used was known as *'the hog's back'*. It consisted of a series of spikes set in a board, through which the sheaf was pulled. But the introduction of the flail saw a solution to the problem which was to last for centuries, and indeed it is known that flails were used in remoter parts of Scotland at the beginning of this century. A flail was made of a staff of wood up to five feet in length and a shorter *souple* or threshing arm with the two joined together by a short length of hide or rope. The threshing was accomplished by placing one line of sheaves side by side on the barn floor and laying a further line on top in the opposite direction in such a way that the heads overlapped. Swinging the flail over his shoulder, the worker would beat his way down the line. The sheaves would then be turned and further flailing would be given.

The first attempts at mechanical threshing were based on a number of flails attached to a rotating axle, but gradually more effective practices evolved, and by the early eighteen-hundreds many threshing mills, capable of screening light oats and extraneous material from the threshed crop, were in operation in Scotland. In due course, one of the great country sights was to arrive on the scene — the travelling mill, pulled, first, by a team of horses and later by an enormous steam traction engine.

While it is sad, from an aesthetic point of view, that today's masterpiece of the combine harvester has removed the sight of stooks and stacks from the countryside, there are few, if any, allied to the land who regret their departure. Harvest work was hard work. Sheaves, when wet, were heavy to handle. Stacks took much time and effort to build, and heating, resulting in a poor parcel of oats for the miller, was frequently a worry. Where horses were kept, the need to have their first meal of the day settled before commencing work meant a four o'clock in the morning start for the farmer or farm servant. But the memories of those old scenes linger still — it is good that they should.

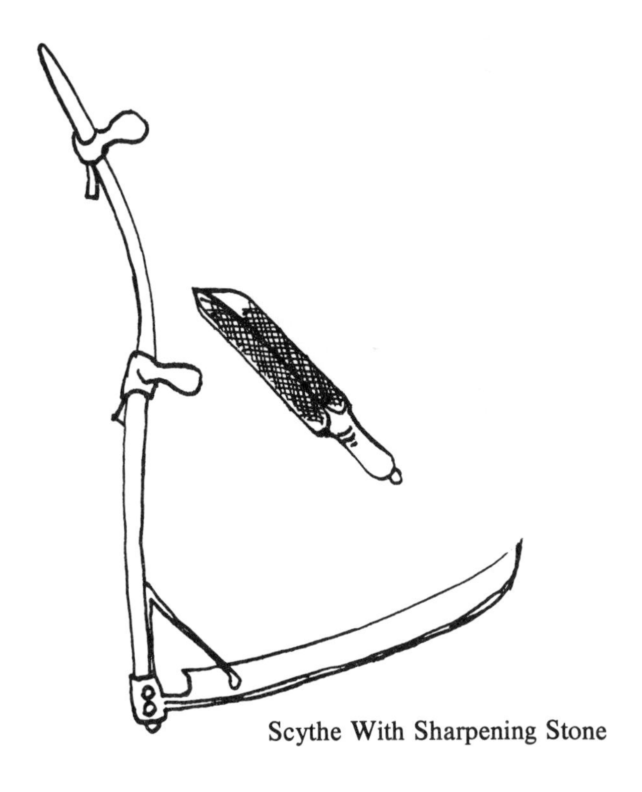

Scythe With Sharpening Stone

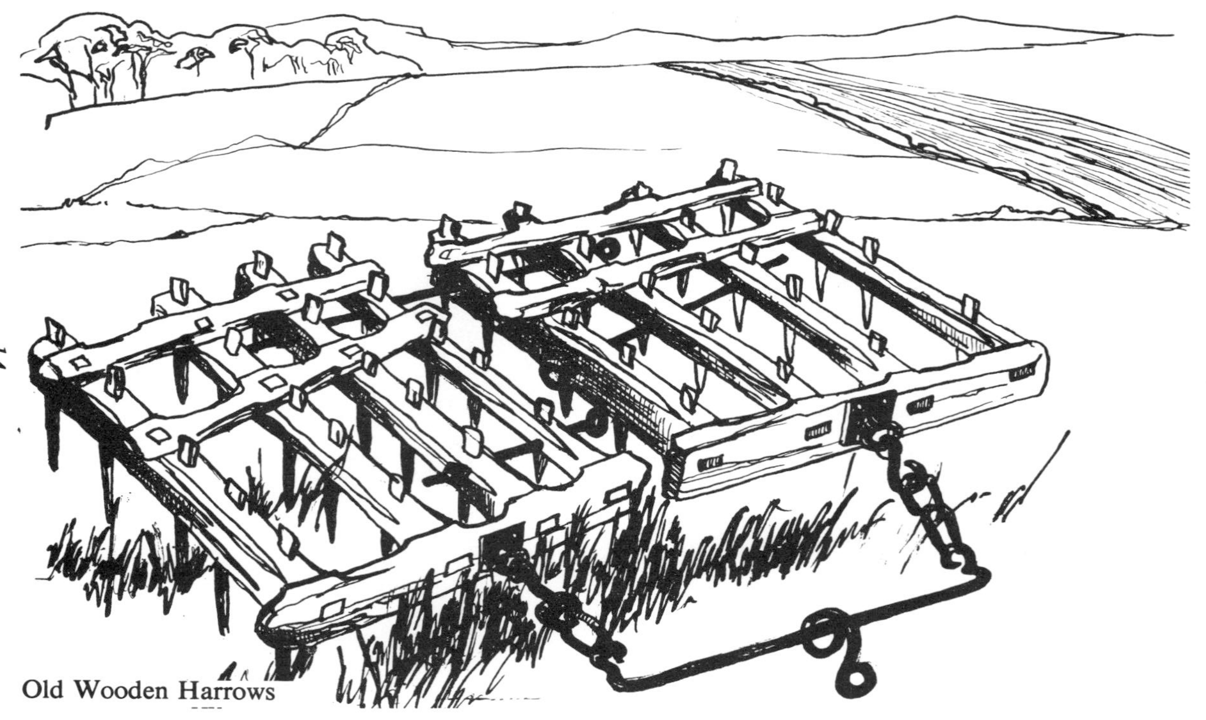

Old Wooden Harrows

CORN RIGS, LEA RIGS AND END RIGS.

'Corn rigs, an' barley rigs,
An' corn rigs are bonie:
I'll ne'er forget that happy night,
Amang the rigs wi' Annie.'

Burns.

'That fruits and herbage may our farm adorn,
And furrowed ridges teem with loaded corn.'

Fergusson

Dry stane dykes are common enough sights in Scotland. It was not always so. In the Middle Ages movement across the land was unhindered by dykes, fences or hedges. Most people lived in villages and the run-rig system of farming was in operation. This meant that ground capable of supporting crops was divided into strips or ridges with ownership or tenancy of the strips being divided amongst different people, so that it was unusual for one man to hold adjoining strips. The strips came into being because of the nature of the cultivations. The absence of field drains encouraged cultivations which raised the level of the crop-bearing ground, leaving land at a lower level, between the ridges, to act as open drains. Even today it is common to see land, especially on hill-sides, showing regular, if gentle, corrugations across fields indicating its past cropping in ridges, or rigs as they came to be called.

The breadth of a rig was normally between five and ten metres; its length depended on the make-up of the soil and the slope of the ground. It is known that the rigs frequently were rotated between different farmers to give more equitable income opportunities. The best land was kept for the growing of crops while the poorer ground was occupied by stock. Arable and grazing ground were thus distinct.

The change from the run-rig system to self-contained farming units was a natural progression encouraged by a number of different circumstances and took many years. More mouths to feed, as the population increased, demanded that more crops be grown, and this was done at the expense of the grazing ground. The opportunity to develop a cattle trade with England required more winter feeding for the stock, and the number of barns and byres had to be increased; these were built near to the best crop-carrying ground so that the manure did not have to be moved so far. Thus came the development of the in-bye (or '*muckit land'*) and the out fields, which although they might be cropped occasionally, were mainly grazed. The introduction of field drains eliminated the need to have land lying uncultivated between ridges, and this ground was gradually brought under the plough. Wheat began to be grown, and, as it required winter sowing as opposed to the spring sowing of oats and barley, the cultivated ground had to be protected by dykes or hedges from livestock which normally would have been feeding on the in-bye ground at that time of year. In time, the fertility of the out-bye land was raised until as much of the farm as possible came into a rotation of crops including grass or lea. Where possible, oats came after lea in a rotation, as no crop does better on ploughed up turf.

When ploughing a field, of course, the furrow could not be run to the edge of the field, as space had to be left for the horses to turn before ploughing again in the reverse direction. It was this space at the end of the field that was known as the end-rig — the last bit of ground to be ploughed.

MILLING

**'Hey the Dusty Miller,
And his dusty coat,
He will win a shilling
Or he spend a groat.'**

Burns.

Some of the earliest references to milling are contained in the Bible where we learn that people ground manna in mills, or beat it in a mortar. In *Judges*, chapter 16, verse 21, it can be read that, after Samson had been taken by the Philistines, *'he did grind in the prison house.'*

Those early methods of milling involved rolling a stone of round or cylindrical shape over a flat or slightly concave stone on which grain had been spread. Such stones, dated around 1500 B.C., have been discovered in Strathclyde. But the big leap forward in milling technology came with the introduction of querns, and these were certainly being used in Scotland when the Romans arrived. Indeed, querns provided such a beautiful and elegant solution to the problem that there was no further improvement for centuries. They were, in fact, still in use in Orkney and Shetland at the beginning of the present century.

A quern was made of two flat, round stones, each about two feet in diameter, the lower stone remaining stationery whilst the upper stone, fitted with a handle, rotated above it on a spindle arrangement. The grain was fed into a hole in the top stone and was milled between the two flat stone surfaces before being directed out to the edges by means of grooves. Initially, the ground meal was collected from a cloth on the ground on which the quern was

Water Mill

placed. In later years, the quern was placed on a stand, allowing the meal to be collected in a hopper arrangement made of skin. As time progressed, larger stones were brought into use, with animals providing the necessary power, and there are indications that around the eleventh century the first water mills were in existence. Regrettably, only a few water mills have been maintained in a working condition, the present fashion being to convert them into luxury homes or hotels. It is surprising that, even in the windier parts of Scotland, windmills do not appear ever to have been common, although a few were known to be operating in Aberdeenshire and in Orkney in the sixteenth and seventeenth centuries. In some parts of the country, attempts were made by forward-thinking pioneers to drive mills by tidal power, but no great success was obtained.

It must be remembered that first attempts at milling did little more than detach the outer shell of the oat from the inner kernel or groat. There was then a need to separate this mixture of shells and groats so that the latter could be ground into meal. One method adopted was to throw the mixture into the air, at a windy place, from a basket known as a *'wecht',* allowing the shells to blow away while the groats fell to the ground and were collected on sheets which had been laid out. This method of separating the groats from the shells gave rise to the name *'Shielhill'*, which is still a fairly common place-name in Scotland. The 'shiel', of course, is the shell from the oats, and the Shielhill is the windy hill where this winnowing process took place.

A later improvement on this inevitably wasteful method of separation was to site the barn in such a way as to make use of the prevailing wind. Doors on opposite sides of the barn would be opened to provide a wind channel and, as the grain was tossed up, the chaff and other light material would be carried away. As mills became more sophisticated, additional stones were added so that one set would first remove the shell and the second set would grind the groats into meal.

It is difficult to mill soft or moist grain, and the first efforts at drying or kilning, as it came to be called, included pot drying over a fire. Gradually home kilns came into existence. These varied very much in size and style throughout the country, but in general were stone built structures with a peat fire at ground level and a flue arrangement allowing heat to rise to a drying floor. Kilns can be traced back to Roman times, and even today, especially in Caithness and the Islands, one can see the ruins of bee-hive shaped kilns on the ends of barns.

As the size and number of stones used in milling increased, so their dressing became more of a craft. It was the job of the stone dresser to cut grooves of the correct size and angle in the stones to ensure that the meal did not choke the mill when it was being ground, and to help it find its way to the outside edge of the stones for collection. At first, stones for milling were obtained from nearby quarries and later the sandstone from the Derbyshire Peak District found favour for the top, revolving, stone. Eventually, though, every commercial mill used '*buhr*' a very hard, flinty stone, imported from France.

For many years in Scotland, the milling of oats was the root cause of friction between tenant farmer and landlord. When building mills, the lairds ensured that the mills would be profitable by entering into agreements with their tenants that they would all dispose of their querns and send their milling oats to the laird's mill for grinding. Additional clauses in the agreement sometimes obliged the tenant to help in the maintenance of the mill and its dam and lade. This obviously was not, and could not become, a long-term happy relationship. It was a custom, also, for payment to be made to the miller in the form of his retention of a proportion of the meal. This led to many arguments over the weight of oats sent to the mill, the apparent loss of weight during milling, the quantity of meal produced, and so on. Even where the tenant did not send his oats to his own laird's mill, he was obliged to make a standing payment, and legal action might be taken against any tenant farmer seeking to avoid payment. These payments, in some cases, had to be made in perpetuity, and with the demise of the country mill have led to more than a few squabbles.

Today's oatmeal mill is a highly complex affair which still demands the craftsman's touch if a high class product is to be obtained. On arrival from the harvest fields, the grain is first weighed, then quality-control checked for such things as colour, size and weight of kernel. It is dried if necessary and subjected to an initial cleaning process to remove straw and other items which may have been picked up during harvesting. Then it is stored in silos capable of holding thousands of tons of oats so that the production of meal may continue unhindered right through the year.

From the silos the grain undergoes a further cleaning to separate out the light and small oats and husks, and this is done by passing the oats over and through a series of sieves and aspirators. More difficult to separate out are weed seeds similar in size and density to the grain, but which, if milled with the grain, would produce unwelcome coloured particles in the meal. Special equipment working on the surface adhesiveness of the weed seeds may be used for this operation.

The kilning of the oats follows. This is probably the most important operation of all, as it is the kilning that brings out the flavour and aroma which make oatmeal products so unique and impossible to substitute with any replacement or artificial product.

On leaving the kiln, the oats are given a further regrading before they pass through the hullers which remove the outer husk of the oat. The naked kernel is now known as a groat, and it is the groat only which passes on to the cutters or stones to produce the texture of the meal required. This may be the chunky pinhead, or coarse, medium or fine oatmeal. Oatflakes are made from the large pinhead oatmeal, which for that purpose is first cooked by steaming before being fed through large, heavy rollers which give the distinctive shape. Cooling and packing then follows.

Perhaps surprisingly, a useful export trade in porridge oats to warm climates exists. It seems that the ubiquitous Scot finds nothing incongruous about eating a hot oatmeal breakfast in the tropics.

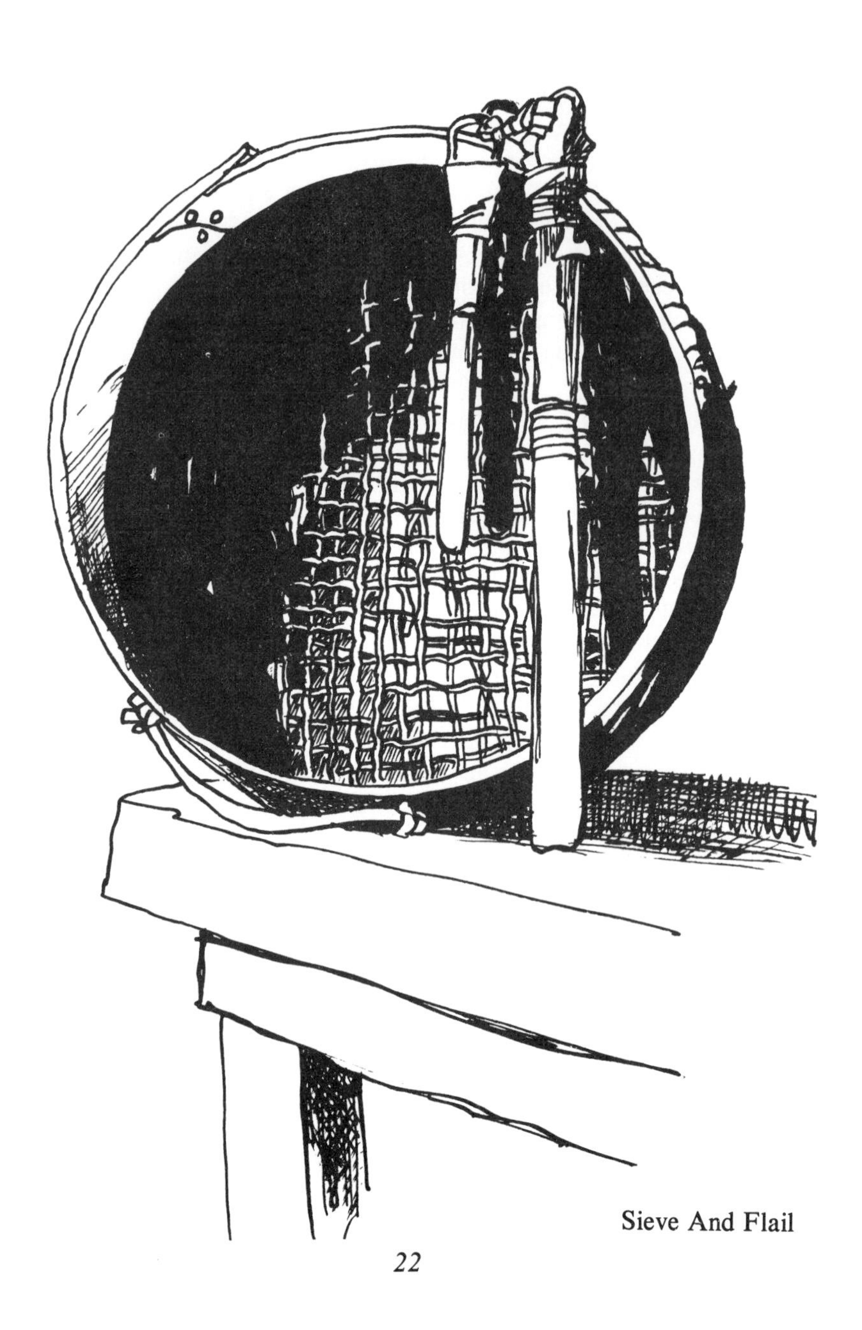

Sieve And Flail

WEIGHTS AND MEASURES

'Nor dribbles o' drink rins through the draff,
Nor pickles o' meal rins through the Mill-e'e.'

Lady Grisell Baillie

'A Guid New-Year I wish thee, Maggie!
Hae, there's a ripp to thy auld baggie!'

Burns

The Scots tongue is rich in terms which, though they may be unknown to the listener, are so descriptive that they immediately call their meaning to mind. A pickle of meal is a case in point; it just *has* to be a small quantity flowing through the fingers. Similarly, the ripp of corn given to the old horse in Burns's lines conjures up the picture of a handful of oats being pulled from a sheaf lying in the barn. Such terms as *'pickle'* and *'ripp'* are a reminder of the old vocabulary of weights and measures that existed long before a metric, or even an Imperial, system was introduced. Many of the old Scots weight and measure terms were the same as those used in England but often the actual weight or measure differed. For example, a Scots peck was about eight pounds (8 lbs.), while the English peck was a pound more than that. Indeed, it seems likely that even within Scotland the weight or measure accorded to a particular term might vary from place to place.

The old way of putting a value on a sample of oats was by its bushel weight. A bushel is a measure of volume, and the original standard weight given to a bushel of oats was forty pounds (40 lbs.), although this was later to be increased to forty-two pounds. The weight of a bushel of oats depended largely on the plumpness of the grain, although it could, of course, be affected by how well light oats, empty husks and so on had been removed from the sample. It was sensible then to decide a value by the bushel weight, as a high bushel weight, say forty-five pounds, would give a greater extract of meal than one of forty pounds. It is worth noting that the bushel weights of oats grown in Scotland are probably the highest obtainable anywhere in the world.

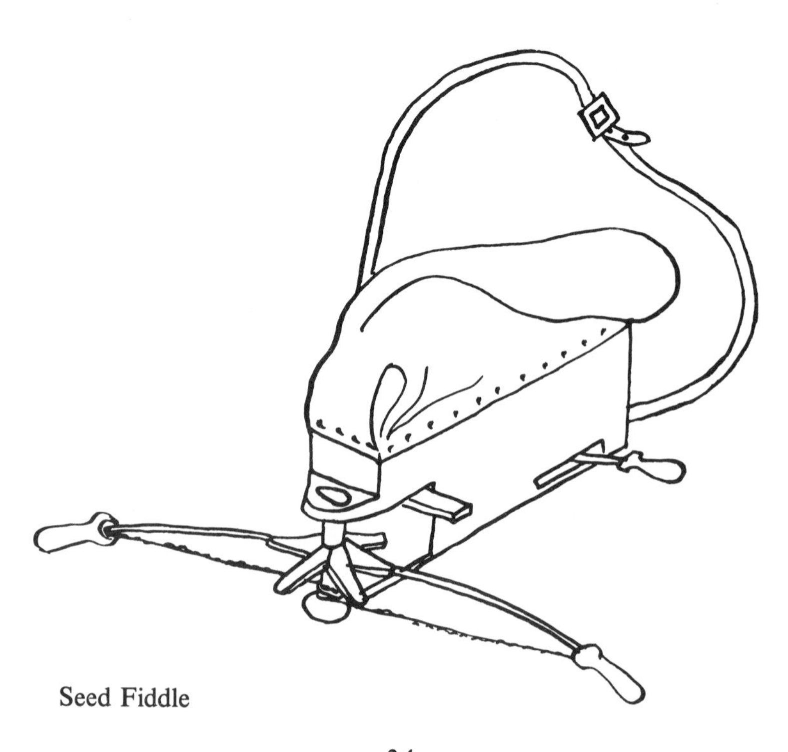

Seed Fiddle

Oats, however, were not traded in such small quantities as bushels, but in a measure known as a 'quarter', which was eight bushels. When the Corn Sales Act was passed in 1921, whereby all oats had to be sold by weight, the quarter became standardised at three hundredweights. Three hundredweights can hardly be lifted by one person and the practice evolved of weighing oats into sacks of a hundredweight and a half, or half a quarter. Bearing in mind the amount of man-handling that had to take place on farms and in mills, it is no wonder that extraordinary muscles developed.

The position was even worse with barley and wheat. Due to their higher bushel weight, half a quarter of barley weighed two hundredweights, and half a quarter of wheat two and a quarter hundredweights. Stacking bags of those weights was a job for a Samson. Having taken receipt of his oats in quarters, the miller would then sell his meal by a measurement known as a '*boll*'. This unit was standardised at a hundred and forty pounds. A typical allowance of meal given to a married farm servant was half a boll a month, while the single man living in the bothy would receive a '*firlot*', half the married man's allowance.

A measure frequently used when dishing out oats for the work horses was the '*lippie*' which weighed a pound and three quarters. In the days when a farm servant was allowed garden ground for growing his kail, he would be given what was known as a '*lippie-sowing*' — an area of ground equivalent to what could be covered by broadcasting a lippie of seed.

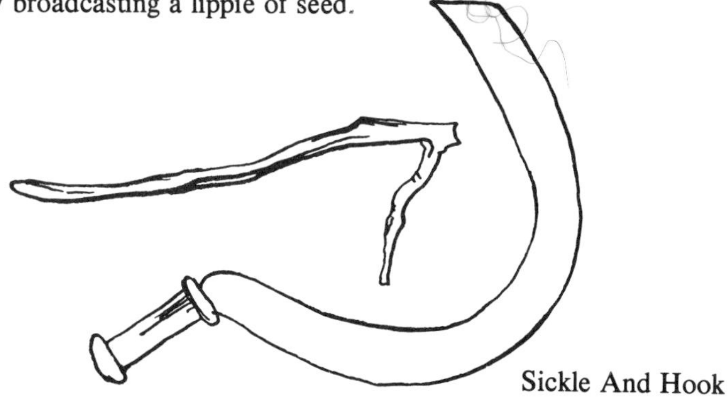

Sickle And Hook

Stackyard

PORRIDGE

'On sicken food has mony a doughty deed
By Caledonia's ancestors been done.'

Fergusson.

In many Scottish households the changing of the clock in October heralds the re-introduction of porridge on to the breakfast table. It is a sensible tradition, for few breakfast dishes give greater fortification against the cold.

The spelling of the word seems to have been settled around the end of the eighteenth century. Prior to that time a variety of spellings were in use: parrage and parritge were common, and literature up to the end of the sixteenth century makes reference to porrage and inevitably porage, the descriptive term used by one major manufacturer today. According to many dictionaries, porridge is a mixture of oatmeal and water which undergoes a boiling and simmering treatment. Salt is a necessary additive and the true Scot deplores its replacement, as is frequently done South of the Border, by sugar.

For hundreds of years porridge was the staple food of the poorer people in Scotland. Personalised as 'they' in many households, the porridge pot was never moved from the fireside and the eating of porridge while standing up was always an acceptable custom. At the table two bowls would be used, one filled with porridge and the other with milk. Rather than milk being poured over the porridge, it was customary for each spoonful to be dipped into the milk bowl before eating. Milk, however, was not necessarily the only accompaniment to porridge. Honey, syrup and treacle were popular and there are records indicating that ale and porter were not without their devotees. Old porridge bowls were normally made of wood, and the spoon from horn, which did not transmit heat to the mouth so readily. The stirring of the porridge pot was done using a long tapered stick, frequently with a carved head, and known, according to locality, as a 'spurtle' or 'thrievel'. The stirring of porridge, of course, is always done in a clock-wise direction to ensure good luck.

It was customary for a farm servant to use two 'kists' or chests. Blankets were kept in one and meal in the other. In order to retain its condition, the meal had to be tightly packed, an operation which involved rolling up the trouser legs and trampling inside the kist with bare feet. In the bothies of the old farm touns where the single men and sometimes, to the chagrin of the old kirk, the single women also, were accomodated, porridge was known to be given a variation that would startle the most fervent oatmeal fanatic of today. Before taking to the fields, the farm servant would ladle some porridge into a drawer of a chest where it would cool off. At night it would be sliced into chunks and, rejoicing under its new name of 'calders', would supplement an egg or fish dish. It is interesting to note in this connection that salmon, so widely considered nowadays to be a luxury food, was once so common that frequently a farmer, when feeing a farm servant for a year, would agree not to provide him with salmon more than so many times a week!

It takes time, of course, to make porridge, and when time was short, the answer was to make brose. The pouring of boiling water over oatmeal and salt was all that was required to prepare this meal, which was then eaten as gruel with milk over it.

Another famous variation of porridge was known as 'sowans' — a dish which Dr. Johnson consumed heartily at Tobermory on his Highland journey.

To make sowans, 'sids' (which are the inner husks of the oats with kernels sticking to them), would be steeped in twice their bulk of water for a week or so until the mixture became sour. The liquid would be drained off and fresh water and salt added to the sediment which was then boiled until it became rich and creamy.

Sometimes the sediment itself would be worked into a dough and rolled in oatmeal. This was a regular school 'piece' for children and, because sowans kept well, was a favourite food when going on a long journey. In the days of farm touns, sowans were often mixed with milk, or, when milk was scarce in the wintertime, mixed with ale and carried to the fields in the inevitable lemonade bottle. Understandably, the eating of all this oatmeal food was claimed by some to produce the skin disease 'Scotch Fiddle', so named because of the continual scratching it promoted, but medical opinion, as ever, was divided on the suggestion.

Porridge, today, is regarded as a first breakfast dish (before the bacon and eggs), and has been considerably sophisticated as a morning cereal. When made from pin-head oatmeal, a period of soaking before long cooking was required, and although the resultant porridge had a beautiful creamy texture, the lumps that inevitably found their way into the plate were not acceptable to the more refined palate. Nearly all porridge today is made from rolled oats which involve a minimum of effort and cooking time, but which, thanks to the continuation of kilning, still retain their unmistakable and mouth-watering flavour.

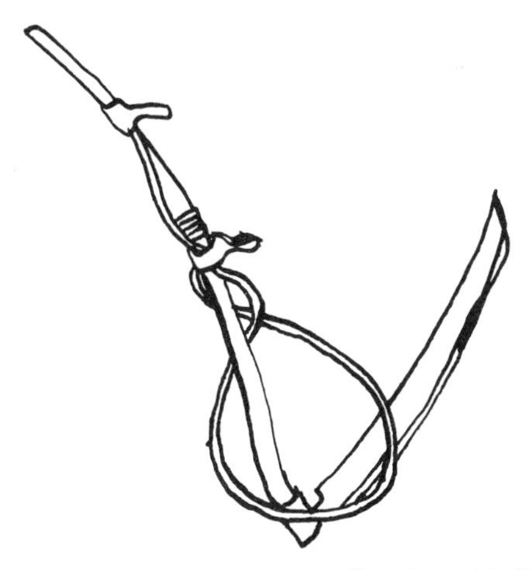

Scythe with Willow Cradle

Housewife Using Spaithe

OATCAKES

'Wi' buttered bannocks now the girdle reeks,
I' the far nook the bowie briskly reams.'

Fergusson

'The best o' bannocks tae yer tea
Gangs doon yer craig like leaves in spate.'

Violet Jacob

The current enthusiasm for health foods is happily increasing the popularity of oatcakes. High in nutritional value, their substantial fibre content provides the balancing roughage to some of today's high cholesterol foods. But whether the accompanying food is "party fayre" or cheese or honey, oatcakes are the perfect partner. They have the ability to go with any food, even ice-cream (an introduction which must be credited to the Americans), although Burns's liking for eating them with warm ale is more understandable. Oatcakes today are exported to all parts of the world, with Canada, no doubt because of its large Scottish immigrant population, being the largest market. But food-conscious France takes its quota, with the Bretons showing a particular partiality for the product.

Well into the present century it was a farmhouse custom for a large mix of oatmeal dough to be prepared weekly and stretched out over a rope to dry, allowing the cook, in due course, to tear off appropriate strips to put on the girdle.

The part of the country one is in determines whether the term 'bannock' is synonomous with 'oatcake'. In western parts the words are interchangeable, whilst in the East and North-East in particular the bannock may contain a proportion of wheat flour, and is an item for the sweeter toothed. Indeed, in those areas, the description of it as being a cross between an oatcake and a digestive biscuit is not a bad one. While the name 'farl' lingers on, the original product, unfortunately, does not. A farl was a large oatcake baked slowly only on one side to bring out the full flavour of the oatmeal. As it cooled, the edges curled up, not a suitable shape for present-day packaging.

But of course all oatcakes do not taste the same, and the connoisseur can point to their subtle variations in much the same way as the whisky drinker contrasts a malt from Islay with one from Speyside. The amount of kiln drying or toasting will bear upon the nuttiness of the flavour; they may be baked to varying degrees of hardness and the supporting ingredients to the oatmeal, such as salt and shortening, and even malt and sugar, may be varied in their proportions. Important, too, is the degree of coarseness of the oatmeal used. In the more northern parts of the country, the demand is for a floury type of oatcake made from finely ground oats which more readily form a dough. In more southern parts the taste is for an oatcake with more bite to it, and pinhead or coarse oatmeal will be the major constituent to provide the necessary crunch.

The traditional shape for an oatcake is round, the three-cornered shape so often seen being a result of quartering a large, circular oatcake. Today, all oatcakes are thin, but old cookery books refer to thick bannocks like the 'snoddie' and the 'mill bannock', the latter being twelve inches in diameter and an inch thick, with a hole in the middle — they certainly must have liked their oatcakes in those days!

In more superstitious times special properties were attributed to oatcakes. A large bannock placed above the door, for instance, was considered a protection against fairies taking away a new-born child, while the remedy for a cow cursed by a witch was to milk her through a hole made in a bannock.

The earliest method of baking oatcakes was to place the dough, made from meal and water, on to a flat baking stone set at the side of the fire. A leap forward in technology, however, came with the introduction of the girdle. A girdle is a thin, round, iron plate, with a semi-circular handle attached to the sides in such a way that it may be suspended over the fire from a hook. The name is derived from the old French word *'gridil'.* The girdle was invented and first made in Culross, Fife. In 1599 James VI granted the burgh the exclusive privilege of manufacture, and this was confirmed by Charles II in 1666. Later, a nearby iron works took over its production. The girdle allowed quicker and more regular baking than was possible with the baking stone, although bannocks, after being lifted from the girdle, were often placed against a hot stone for a final hardening and toasting.

A variation of the girdle was the 'brander', where the flat surface was replaced by a series of wavy iron bars. The brander was used when the oatmeal was mixed with the meal from other crops such as beans, rye or barley so that a stiffer dough was made.

Associated with the girdle were the scored or notched bannock stick for rolling out the meal mixture, and, of course, the beautiful, heart-shaped instrument known by a variety of local names such as 'spathe', 'spade', or 'spurtle', which was used to lift the hot oatcakes from the girdle. The girdle was never washed but would be cleaned when hot with coarse salt and a cloth or paper. It is interesting to note that spathes were seldom used in latter day bakehouses, because the bakery girls, eschewing the use of an instrument, had become adept at lifting the thin baked oatcakes from large hot plates with their fingers without suffering burns.

In an age when so many of our foods seem to have lost their authentic flavours, it is nice to be able to acknowledge that the large scale Scottish bakers specialising in oatcakes are producing a product worthy of the name — a tribute to the marriage of new baking technology and simple, honest, unadulterated ingredients.

Haggis

HAGGIS

'Fair fa' your honest, sonsie face,
Great Chieftain o' the Puddin-race!

Burns

'He dwelt far up a Heelant glen
Where the foamin' flood an' the crag is,
He dined each day on the usquebae
An' he washed it doon wi' haggis.'

David Rorie

Alone amongst the oatmeal foods, haggis presents a caricature image. Stories of haggis-hunting abound. It is difficult to see why the haggis should provoke such mirth. In every way it is a superb sausage, and if its skin should be a beast's paunch instead of its intestine (as encloses its continental cousins), this merely emphasises the uniqueness of the haggis.

The haggis has a proud history. The Ancient Greeks as well as the Romans knew its pleasures: Dunbar mentions it is his fourteenth century writings: Dr. Johnson feasted on it during his Scottish tour. Queen Victoria, in her diaries, records her enjoyment at eating haggis at Blair Castle during one of her journeys. No wonder Burns queried the right of anyone to regard it *'wi' sneering, scornfu' view.'*

The word 'haggis' seems to have been derived from the French '*hacher*' — to chop up or mangle. A French derivation would be appropriate for this old dish, as up to the Union of the Crowns in 1603, French influence was strong in Scotland. In the days of the Auld Alliance commercial links between the two countries were close, French troops were stationed in Scotland, claret was the drink of the well-to-do, and it is not difficult to imagine some influence from that gastronomic land infiltrating Scottish kitchen habits. Indeed, there are proofs that this did happen. In Scotland, a leg of lamb or mutton is invariably called a '*gigot*', which of course is the French for 'leg'. And a serving dish is an '*ashet*', which certainly derives from the French '*assiette*' or plate. There are many other examples and not only in the culinary field.

Burns, in his 'Address to the Haggis', was not afraid to air his knowledge of French cuisine, even though it is fairly clear that he thought but little of it.

'Is there that owre his French *ragout,*
Or *olio* that wad staw a sow,
Or *fricassee* wad make her spew
Wi perfect sconner
Looks down wi' sneering, scornfu' view
On sic a dinner?'

After the collapse of the '45 rebellion, many Scots fled to France, taking their love of haggis with them. It still holds recognition in that food-conscious country today as '*Puding de St. Andre.*'

In terms of taste, haggis is supreme. It is only when some thought is given to its contents that an odd eyebrow may be raised. Yet heart and liver are good foods, and the oatmeal, onion and suet mixture has sustained the Scottish appetite for ages with dishes like mealy puddings and skirly-in-the-pan. There have been a few variations on the basic mix. At one time a less coarse haggis made of sheeps' tongues and kidneys was well liked. Highland chiefs were reputed to banquet on haggis made with venison, while Haggis Royal contained mutton, vegetables and red wine.

But always, oatmeal was a major ingredient; and, as has long been recognised, the best accompaniment to haggis is clapshot, a mixture of mashed potatoes and turnip with dripping and suet.

Burns's lines, in his 'Address to a Haggis,' (*Then, horn for horn, they stretch an' strive, De'il tak the hindmost, on they drive.'),* reminds us of the old custom of placing a bowl in the centre of the table for communal supping, when the youngsters, armed with long horn spoons, had to eat fast if their stomachs were not to remain empty. And today, as in Burns's time, haggis is a democratic dish, being enjoyed at all levels of society.

It is no longer common to see a sheep's pluck boiling in the kitchen (with the wind-pipe hanging over the side of the pot to let out the impurities) ready for the making of a haggis. But the steady guzzling of haggis, rising to a consumption peak at Burns Supper time, shows that here indeed is a dish that stands *'Abune them a'.'*

Steam Threshing Rig

Scything Oats

ATHOLE BROSE

**'Aye since he wore the tartan trews,
He dearly lo'ed the Athole Brose.'**

Attributed to Neil Gow

Oatmeal does not readily suggest itself as the basis of an alchoholic drink. Athole Brose, though, is something different, and has its following with many sophisticates throughout the world as the ideal drink with which to start a party. Doubtless the occasion will have a Scots overtone, although Athole Brose is no tartan gimmick; it is a drink which has stood the test of time.

The ingredients of Athole Brose are oatmeal, honey and whisky. For medicinal purposes an egg may be added, and with the growing custom of serving it as a sweet at the end of the meal has come acceptance that cream may be included.

In her *'Leaves from the Journal of Our Life in the Highlands'*, Queen Victoria records how she and Prince Albert drank Athole Brose when they visited Blair Castle in the September of eighteen forty-four. Many thousands of people visit that most beautiful castle at Blair Atholl every year, and it is a fair guess that most of them will indicate interest in the famous drink.

While the origins of Athole Brose are obscure, its evolution can be understood. With the Highlanders' staple food of oatmeal and his own distilled spirit being carried on his travels, it is not difficult to imagine some simple experimentation with additives, and the cry of acclamation when honey was found to be the perfect partner to the oatmeal and whisky.

There is, however, one legend which may explain how the drink came to be associated with the House of Atholl. In the fifteenth century, Iain McDonald, Earl of Ross and Lord of the Isles, was widely regarded as a dangerous menace to society, and his capture was sought by many, including the then Earl of Atholl. It came to the notice of the Earl that McDonald frequently drank from a particular well, and the Earl gave instructions that the well water should be laced with whisky and honey. In due course McDonald came to the well and, understandably, finding the water very much to his liking, dallied too long, becoming an easy prey for Atholl's men.

Athole Brose gained a high reputation as a medicine. There is a story from the Atholl Estates, dated around 1800, of the daughter of an inhabitant of Atholl being placed in one of the first boarding schools in Edinburgh, where she was seized with a violent fever. Her father was sent for, as she was considered to be on the point of death. On his arrival he was told that everything the physicians could do for her had been done, but without effect. 'Has she had any Athole Brose?' he asked. On receiving a negative reply, he had a good dose of it instantly prepared and had the child swallow it, and then happily watched her almost immediate recovery.

While proprietary brands of Athole Brose are on the market, most regular users make up their own batches so as to produce a drink specially pleasing to their palate. The consumption of Athole Brose peaks at Burns time and St. Andrew's Night, and many a Scot returned from exile has regaled his friends with hair-raising stories about preparing quantities of the Brose in uncommon utensils in jungle or desert climes. Sir Robert Bruce Lockhart in his fine book 'Scotch', tells how he organised a St. Andrew's Day dinner in the Czechoslovakian capital of Prague. Unfortunately he delegated the making of the Brose to a Sassenach who, in an attempt to make his mark, laced it with slivovice, the potent plum brandy. Sir Robert's comment that *'He suffered for his intervention in Scottish affairs'* is, one can imagine, a masterpiece of understatement.

Broadcasting Seed (c. 1890)

Winnowing Oats in Shetland (c. 1900)

Grinding Oats in a Quern (c.1900)

Bothy Lads in Angus

Cooking Implements:-
Kail Pot
Girdle
Spurtle
Spaithe
Brander
Bake Stone

Bakehouse of the old days producing Oatcakes

Steam Threshing Rig, 1908

Delivery Vans of an earlier day

CREAM-CROWDIE or CRANNACHAN

'Ance crowdie, twice crowdie,
Three times crowdie in a day:
Gin ye crowdie ony mair,
Ye'll crowdie a' my meal away.'

Burns

The crowdie Burns was referring to in his lines differs considerably from the elegant sweet which today bears the Lowland, double-barrelled name of cream-crowdie, or the Highland, Gaelic name of crannachan. Burns was using the word 'crowdie' to indicate a porridge dish, in much the same way as, in 'The Holy Fair', he uses the term 'crowdie-time' to indicate breakfast. The word, however, has a different connotation in the Highlands, where a kind of soft cheese is known as crowdie. It was the Gaelic word '*crannachan',* meaning half-churned cream, which came to be regarded in the Highlands as the equivalent of cream-crowdie and has become the predominant name given to the sweet today.

Crannachan was, originally, a mixture of toasted oatmeal, cream and sugar prepared for festive occasions, especially Hallowe'en, when small charms would be placed in it in the same way that coins and trinkets are secreted in Christmas puddings today. Crannachan has always been a dish without a precise recipe, it being the custom for the lady of the house to make it according to her guidman's taste. This, we may take it, explains the addition of whisky, which further expanded its popularity. But there was to be one further addition which would ensure its entry into gourmet circles.

The counties of Angus and Perthshire produce over eighty per cent of all the raspberries grown in Britain. The light loam encourages the free surface rooting necessary for the establishment and growth of healthy plants, and the moist climate ensures plump and succulent berries. In this area, fields of up to thirty acres of raspberry canes may be seen, and there are few gardens without their row or two of rasps. With such a crop in superabundance it was inevitable that housewives, besides making substantial quantities of jam, would find another use for their berries. Raspberries and cream have always been a popular treat in this area, which is

Abandoned Steam Thresher

also renowned for the quality of its oats. The population is particularly sweet-toothed, a fact readily confirmed by the number of confectionery manufacturers who originated in the area and by the common practice of taking a dram with lemonade instead of water. The experiment of blending oatmeal, cream, whisky and raspberries is not difficult to imagine, and thus crannachan arrived on the scene as a sweet quite without equal.

Not that crowdie and cream itself is to be despised in any way. Far from that! The crowdie is nearly enough what is often today called cottage cheese, and indeed it originated in cottages as a staple food for peasant and crofter. It was simple enough to make — after the milk from the house cow had been skimmed and the cream used for butter making, the skim milk was kept for a week or so until it soured and thickened. It was then heated slowly until the crowdie floated to the surface, leaving the lovely, acid whey behind. (The whey itself was a fine, thirst-quenching drink, ideal for hay-making or harvesting: nothing was wasted in a crofting household.)

Some of the crowdie was pressed into moulds and kept as hard cheese for the winter, and fine stuff it was.

The great treat, though, was to have crowdie mixed with fresh cream and piled on an oatcake with fresh salted butter. Then you had a royal feast of flavours — acid, sweet and salt, and, better perhaps, a royal mixture of textures, soft, crisp and crunchy.

And every part of that great delight had been produced on the croft. It owed nothing to shops or factories. And that paean of praise must be written in the past tense, for who knows where such delights can be found today.

RECIPES

Scottish cooking is very rich in recipes using oatmeal as a staple ingredient. In this small book it is possible to give only a small selection, and I have chosen those which I can personally vouch for, since they still appear on my table with great regularity, and are eaten with the greatest pleasure by family and guests alike.

OATMEAL SOUP

An unusual, elegant and economical soup suitable for many occasions.

2 oz. (50 g) Rolled Oats
2 medium onions chopped
1 large carrot, grated
2 tablespoons melted butter
1 pint chicken stock
1 pint milk
Chopped parsley
Salt and pepper
Cream for garnishing (optional)

Melt butter in a large pan over low heat: add onions and carrot and cook gently for about 6 minutes: add oats to pan and continue cooking for about four minutes, stirring frequently: add the stock little by little and season to taste: bring to the boil and simmer with lid on pan for 25 to 30 minutes: add parsley: add milk and heat through but do not boil: garnish with cream if desired.

ATHOLE BROSE —— as an aperitif.

Soak a quantity of pinhead oatmeal in water overnight: drain off the liquor and add honey to taste: add an equal amount of whisky, and serve with some of the oatmeal in each glass.

ATHOLE BROSE —— as a dessert

To a lightly toasted tablespoon of pinhead oatmeal add 1 tablespoon of honey and 2 tablespoons of whisky: mix well: gently fold this mixture into a pint of stiffly beaten cream.

CRANNACHAN

Surely this must be the most magnificent of all sweets!

3 oz (75 g) quick-cooking oats
½ pint whipping cream
1 teaspoon honey
Whisky or Drambuie
¼ lb (100 g) raspberries

Toast the oats gently over a low heat, stirring frequently, for about three minutes until lightly browned: beat the cream until it forms peaks: gradually add the honey and lightly fold in the toasted oats and berries: sprinkle the mixture with the whisky or Drambuie: eat,and marvel.

BROONIE

It can only be the natural reticence of the Orcadian that has inhibited the promotion of this most mouth-watering tea-time delicacy.

5 oz (125 g) plain flour
5 oz (125 g) quick cooking oats
6 oz (150 g) dark brown sugar
6 tablespoons treacle
4 oz (100 g) butter
½ teaspoon salt
3 teaspoons ground ginger
2 teaspoons baking powder
1 egg
½ pint milk

Heat treacle, sugar and butter over low heat until butter is melted: mix sieved flour, baking powder, ginger, salt and oats in a large bowl: beat egg in a separate bowl,add milk to beaten egg: add egg and milk mixture to treacle mixture: add the resultant mixture gradually to the flour mixture and mix thoroughly: if you are not by that time too mixed up, bake in a moderate oven (180 C, Gas Mark 4) for about 1 hour.

MUESLI

8 oz (200 g) Rolled Oats
2 oz (50 g) Raisins
2 oz (50 g) Sultanas
1 oz (25 g) Soft brown sugar
1 oz (25 g) Chopped peanuts

Mix all ingredients well and serve topped with honey or plain or flavoured yoghurt.

SKIRLIE

Mealy pudding without the skin, and delicious with fish, mince or 'Chappit tatties'.

2 oz (50 g) butter or roast beef dripping
1 finely chopped onion
Rolled or quickcooking oats
Seasoning

Melt butter or dripping over medium heat and and add onion: Stir until onion is lightly browned: add oatmeal slowly until all the fat is absorbed and the mixture is fairly firm: add seasoning and keep stirring for about 8 minutes.

GRUEL

For countless years gruel has been the great Scottish stand-by for helping the sick return to health.

Fine oatmeal
Sugar
Honey
Salt

Steep some fine oatmeal in water for three or four hours: stir and strain: combine strained liquid with ½ pint of water and bring to boil: simmer for 1 hour: add salt and sugar and a dessertspoon of honey.

OATMEAL STUFFING

A traditional Scottish stuffing for poultry or game.

8 oz(200 g)Rolled Oats
1 onion finely chopped
4 oz (100 g) shredded suet
Seasoning

Mix all ingredients in a bowl, and stuff poultry or game in the usual way.

HERRING IN OATMEAL

Filleted herring
Seasoning
Oatmeal
Lard or dripping

Season herring fillets and toss in oats until thoroughly covered: fry in hot fat, turning occasionally, for about 7 minutes: serve with a mustard sauce.

HAGGIS

Wash the stomach bag of a sheep in cold water: scrape it and soak in cold salted water; wash the sheep's heart, liver and lights and boil these for 2 hours, letting the windpipe hang over the side of the pan: mince the liver, heart and lights, and add ½ lb grated suet and 2 finely chopped onions: toast 1 lb pinhead oatmeal in the oven: mix everything with a very good dash of salt and pepper: fill stomach bag three-quarters full (to allow room for the oatmeal to swell): sew up bag and prick it: cook in a large saucepan of water for 3 hours (making sure water covers the haggis all the time). Serve with 'chappit tatties' and 'bashed neeps' (to the uninitiated, mashed potatoes and turnips).

OATCAKES

1 lb (400 g) oatmeal
2 oz (50 g) butter
1 cup boiling water
1 pinch baking soda
1 pinch salt

Mix all ingredients together and roll out thin: cut into rounds and bake in a moderate to hot oven until crisp. (Alternatively, of course, use the girdle if you are lucky enough to have one. The hotplate of a Rayburn or Aga is a good substitute.)

SWEET BANNOCKS

7 oz (175 g) Rolled Oats
4 oz(100 g) S.R. flour
2oz (50 g) lard
1 oz (25 g) margarine
4 oz (100 g) sugar

Combine oatmeal, flour and sugar in a bowl: melt lard and margarine with 4 tablespoons of water in a pan and pour on the oatmeal, flour and sugar mixture: mix well, roll out and cut into small rounds: place on a baking tray and bake for ¼ hour at 400 F (Gas Mark 6): turn oven down to minimum heat and leave for a further 10 minutes.

MEALIE CANDY

3½lbs (1½ kilos) sugar
1 lb (400 g) treacle
8 oz (200 g) toasted oatmeal
2 oz (50 g) ground ginger

Put sugar, treacle and 1¼ pints of water in a pan and bring to the boil: boil for ten minutes and remove the pan from the heat: with a wooden spoon, rub the syrup gently against the side of the pan until creamy; stir in the oatmeal and ginger: pour the mixture into tins lined with oiled paper and cut into pieces when cool.

BLENSHAW (Blanche Eau)

Oatmeal
Sugar
Milk
Water
Nutmeg

Put a teaspoon of oatmeal into a tumbler with the same amount of sugar. Pour in half a gill of milk and stir until creamy. Pour on boiling water to fill glass. Grate some nutmeg over the mixture and drink when cool.

Binder At Work

SOME FURTHER QUOTATIONS CONCERNING OATMEAL

Scottish literature is full of references to oatmeal and its many uses and benefits. While researching for this book, I have come across many which deserve notice, but do not fit into the body of the book. Rather than lose them again, I have collected them into this separate section.

What the Auld Fowk are Thinkin

'Whan the auld fowk sit quaiet at the root o' a stook,
I' the sunlicht their washt een blinterin and blinkin,
Fowk scythin, or bin'in, or shearin wi' heuk
Carena a strae what the auld fowk are thinkin.
George Macdonald (1824-1905)

Athol Brose

'Charm'd with a drink which Highlanders compose,
A German traveller exclaimed with glee, —
"POTZTAUSEND! sare, if dis is Athol Brose,
How goot der Athol Boetry must be!"'
Thomas Hood (1799-1845)

Haggis and Oatcakes

'Now we are upon the article of cookery, I must own, some of their dishes are savoury, and even delicate; but I am not yet Scotchman enough to relish their singed sheep's head and haggis, which were provided, at our request, one day at Mr. Mitchelson's, where we dined. The first put me in mind of the history of Congo, in which I had read of negroes' heads sold publicly in the markets; the last, being a mess of minced lights, livers, suet, oatmeal, onions, and pepper, enclosed in a sheep's stomach, had a very sudden effect upon mine, and the delicate Mrs. Tabby changed colour; when the cause of our disgust was instantaneously removed at the nod of our entertainer. The Scotch in general are attached to this composition with a sort of national fondness, as well as to their oatmeal bread; which is presented at every table, in thin triangular cakes, baked upon a plate of iron, called a girdle; and these many of the natives, even in the higher ranks of life, prefer to wheaten bread, which they have here in perfection.'

Tobias Smollett (1721-71)
'Humphrey Clinker'

'But first we'll tak a turn up to the height,
And see gif all our flocks be feeding right;
By that time, bannocks and a shave of cheese
Will make a breakfast that a laird might please.'

Allan Ramsey (1682-1758)
'The Gentle Shepherd'

'For here yestreen I brewed a bow of maut,
Yestreen I slew twa wethers prime and fat.
A furlet of good cakes, my Elspa beuk,
And a large ham hangs reesting in the neuk.
I saw mysel', or I cam o'er the loan,
Our muckle pot that scads the whey, put on,
A mutton-bouk to boil, and ane we'll roast;
And on the haggies Elspa spares nae cost.
Small are they shorn, and she can mix fu' nice
The gusty ingans wi' a curn of spice;
Fat are the puddings, — heads and feet weel sung.'

Alan Ramsay (1682-1758)
'The Gentle Shepherd'

'Imprimis, then, a haggis fat,
Weel tottled in a seethin' pat,
Wi' spice and ingans weel ca'd through,
Had help'd to gust the stirrah's mou',
And placed itsel' in truncher clean
Before the gilpy's glowrin' een.

Fergusson

The Miller's Kitchen

'Behind the door a bag o' meal,
And in the kist was plenty
Of good hard cakes his mither bakes,
And bannocks were na scanty.'

Sir John Clerk of Penicuik (1684-1755)

Duncan Dhu made haste to bring out the pair of pipes that was his principal possession, and to set before his guests a mutton-ham and a bottle of that drink which they call Athole Brose, and which is made of old whisky, oatmeal, strained honey, and sweet cream, slowly beaten together in the right order and proportion. Maclaren pressed them to taste his mutton-ham and the 'wife's brose', reminding them the wife was out of Athole and had a name far and wide for her skill in that confection.

Robert Louis Stevenson (1815-1894)
'Kidnapped'

Queen Victoria's Highland Journals

Sept 11 — Wednesday 1844

We passed the point of Logierait, where are the remains of an ancient castle, — the old Regality Court of the Dukes of Athole. At Moulinearn we tasted some of the Athole Brose, which was brought to the carriage.

October 13 — Thursday 1865

The Duchess has a very good cook, a Scotchwoman, and I thought how dear Albert would have liked it all. He always said things tasted better in smaller houses. There were several Scotch dishes, two soups, and the celebrated 'haggis', which I tried last night, and really liked very much. The Duchess was delighted at my taking it.

GLOSSARY

A', *all.*
ABUNE, *above.*
AITS, *oats.*
ANCE, *once.*
ANE, *one.*
AULD, *old.*

BAGGIE, *the belly, stomach.*
BEUK, *bake.*
BIN'IN, *binding.*
BLINTERIN, *unseeing.*
BLYTHER, *more cheerful.*
BOW, *boll, a measure of 140lbs.*
BOWIE, *a milk pail, a dish.*

CA'D, *driven.*
CRAIG, *the throat.*

DOON, *down,*
DOUGHTY, *valiant.*
DRAFF, *wet barley husks.*
DRIBBLES, *succession of drops, gentle flow of water.*
DWALT, *dwelt.*

EEN, *eyes.*
EKE, *to join.*

FAIR FA', *welcome, good luck.*
FALLOW, *a fellow.*
FOWK, *folk.*
FRICASSE, *minced meat in gravy.*
FU', *full.*
FURLOT, *firlot, quarter of a boll i.e. thirty five pounds weight.*

GAB, *the mouth, to speak pertly.*
GANG, *to go.*
GEAR, *riches, goods of any kind.*
GIF, *if, whether.*
GILPY, *a girl.*
GIN, *if, before, until.*
GLOWRIN, *staring angrily.*
GROAT, *silver coin valued at around four old pence.*
GROATIS, *groats.*
GUST, *to taste.*

HAMEIL, *domestic, homely.*
HEELANT, *Highland.*

IMPRIMIS, *firstly.*
INGAN, *an onion.*

KIST, *chest.*

LOAN, *Lane, farm road.*

MAIR, *more.*
MAUT, *malt.*
MILL E'E, *the mill eye, the opening in the top of a grinding stone.*
MOU', *the mouth.*
MUCKLE, *much, large.*
MUTTON-BOUK, *whole body of a sheep.*

NA, *not.*
NAE, *no.*
NEUK, *nook, corner.*

OLIO, *a stew.*
ONY, *any.*
OWRE, *over.*

PAT, *pot.*
PEERIE, *a spinning top, anything shaped like a spinning top.*
PICKLE, *a small quantity.*

RAGOUT, *well-spiced meat.*
RAX, *to stretch.*
REAM, *to foam, to cream.*
REEK, *to smoke.*
REET, *the bottom, the foot.*

SCAD, *to scald.*
SCONNER, *disgust.*
SHAVE - SHAVER, *a funny fellow.*
SIC, *such.*
SICKEN, *such.*
SONSIE, *pleasant, comely.*
SOUPLE, *flexible, swift.*
SPEW, *to vomit.*
STAW, *a surfeit.*
STEEVE, *firm.*
STIRRAH, *a man.*
STRAE, *straw.*
SUBCHARGE, *something additional.*

TAE, *to.*
TENT, *to take heed.*
TRUNCHER, *a trencher, a platter.*
TWA, *two.*

WAD, *would.*
WAME, *the womb, the belly.*
WASHT, *tired.*
WECHT, *sieve-like tool for winnowing corn.*
WEEL, *well.*
WETHERS, *castrated rams.*
WI', *with.*

YESTREEN, *yesterday.*

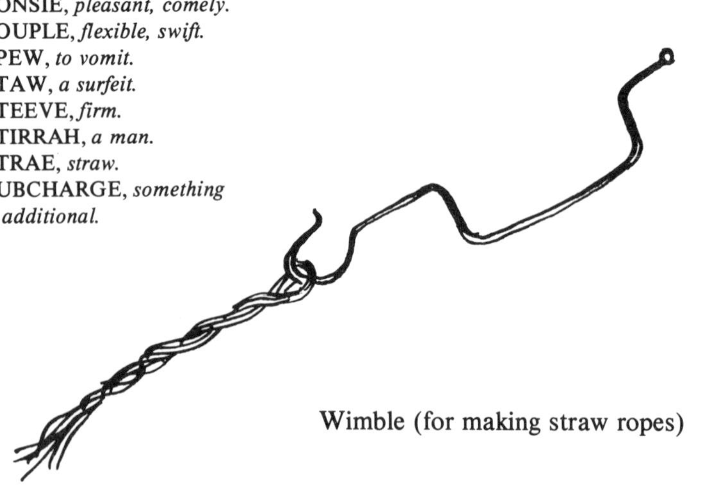

Wimble (for making straw ropes)

Binder